# GROWING MARIJUANA FOR BEGINNERS

## CANNABIS CULTIVATION INDOORS & OUTDOORS

**ClydeBank ALTERNATIVE**

Copyright 2014 by ClydeBank Media - All Rights Reserved.

This document is geared towards providing exact and reliable information in regards to the topic and issue covered. The publication is sold with the idea that the publisher is not required to render accounting, officially permitted, or otherwise, qualified services. If advice is necessary, legal or professional, a practiced individual in the profession should be ordered.

From a Declaration of Principles which was accepted and approved equally by a Committee of the American Bar Association and a Committee of Publishers and Associations. In no way is it legal to reproduce, duplicate, or transmit any part of this document in either electronic means or in printed format. Recording of this publication is strictly prohibited and any storage of this document is not allowed unless with written permission from the publisher.

The information provided herein is stated to be truthful and consistent, in that any liability, in terms of inattention or otherwise, by any usage or abuse of any policies, processes, or directions contained within is the solitary and utter responsibility of the recipient reader. Under no circumstances will any legal responsibility or blame be held against the publisher for any reparation, damages, or monetary loss due to the information herein, either directly or indirectly. Respective authors own all copyrights not held by the publisher. The information herein is offered for informational purposes solely, and is universal as so. The presentation of the information is without contract or any type of guarantee assurance.

**Trademarks**: All other trademarks are the property of their respective owners. The trademarks that are used are without any consent, and the publication of the trademark is without permission or backing by the trademark owner. All trademarks and brands within this book are for clarifying purposes only and are owned by the owners themselves, not affiliated with this document.

ClydeBank Media LLC is not associated with any organization, product or service discussed in this book. The publisher has made every effort to ensure that the information presented in this book was accurate at time of publication. All precautions have been taken in the preparation of this book. The publisher, author, editor and designer assume no responsibility for any loss, damage, or disruption caused by errors or omissions from this book, whether such errors or omissions result from negligence, accident, or any other cause.

Cover Illustration and Design: Katie Poorman, Copyright © 2015 by ClydeBank Media LLC
Interior Design: Katie Poorman, Copyright © 2015 by ClydeBank Media LLC

ClydeBank Media LLC
P.O Box 6561
Albany, NY 12206

Printed in the United States of America

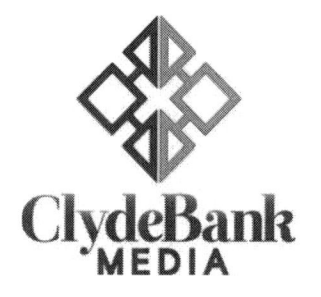

ClydeBank
MEDIA

Copyright © 2014
ClydeBank Media
www.clydebankmedia.com
All Rights Reserved

ISBN-13 : 978-1500555467

# CONTENTS

| | | |
|---|---|---|
| 6 - 7 | | **INTRODUCTION** |
| 8 - 10 | \| 1 \| | Before We Get Started |
| 11 - 18 | \| 2 \| | Knowing Your Growing |
| 19 - 24 | \| 3 \| | Finding a Suitable Location |
| 25 - 28 | \| 4 \| | Concealment Do's & Do Not's |
| 29 - 43 | \| 5 \| | Filling Your Tool Belt |
| 44 - 48 | \| 6 \| | Strain Selection |
| 49 - 54 | \| 7 \| | Growing & Caring for Your Cannabis |
| 55 | | **CONCLUSION** |
| 56 - 57 | | **APPENDIX** |
| 58 | | **ABOUT CLYDEBANK ALTERNATIVE** |
| 59 | | **MORE BOOKS BY CLYDEBANK ALTERNATIVE** |

# INTRODUCTION

Interest in marijuana is at an all-time high. That wasn't a pun, rather it describes the multitude of ways in which legislators, citizens, the media, and public opinion have converged to paint a variety of portraits of the ways in which marijuana should be restricted, used, grown, and sold. Because many of these opinions are conflicting, a vast controversy has developed around marijuana and those who grow it.

The writing is on the wall so to speak; groundbreaking legislation has been cropping up in many states to decriminalize many aspects of the use and transport of marijuana for many individuals. Public opinion and perceptions are shifting away from the status quo, and as the restrictions on both recreational use and medicinal use loosen, so too will demand continue to skyrocket. With what many see as complete legalization on the horizon, the business prospects for both large corporations and small operations are more lucrative than ever before, prompting an industry that is on the edge of its seat for the final legal barriers to be removed and the money to start pouring in. Some experts agree that in time, the consumption of marijuana could rival the demand for other recreational substances such as tobacco or alcohol.

With all of that being said, it's no wonder that more people are now starting to invest time and money into creating their own private operations. The development of a scalable operation has the potential to make a large number of people rich, especially if they are ahead of the game when legalization becomes a reality.

The times they are a changin'.

## What You Will Learn

This is a book for beginners, appropriate for even those readers with no experience whatsoever. In the following pages you will learn how to begin growing cannabis from the ins and outs of growing, to selecting your strain, optimizing your yield, choosing the perfect location, and maintaining your plants. In these pages is everything you need to get started. The concepts found in this book not only apply to those who are looking to grow marijuana solely for personal consumption but to those who are interested in starting their own wholesale operations. Please take a look at the following disclaimer before reading on about the wonderful world of cannabis.

## Disclaimer

The author and the publisher of this book both strongly advise that prior to embarking on your journey cultivating marijuana, you review your region's legislation and understand the repercussions of failure to comply. The author of this book is not a lawyer and cannot provide you with legal advice. No part of this publication should be considered as such. It would be exhaustive to accurately summarize the laws and regulations for each state and locality here. The responsibility of understanding and complying with established laws falls solely on your shoulders. The author of this publication or its publisher is not responsible for any legal repercussions that may result from the growth, consumption, or sale of marijuana.

Now that the legal department has had its say, I would like to thank you for purchasing this book. I hope you enjoy it, and I hope you enjoy the rewarding journey ahead.

Let's get started.

# CHAPTER ONE
## BEFORE WE GET STARTED

First let's clear up any ambiguity that surrounds some of the terms with which marijuana growers will need to be familiar and take a brief look at the history of the cannabis plant throughout the years. Consider it product knowledge for your up-and-coming cash crop.

## Marijuana, Cannabis & Hemp

Throughout this book, and throughout your own research, you will see the three terms marijuana, cannabis, and hemp used frequently. In casual conversation the three terms are often interchangeable, and in some cases this is acceptable, but the difference is important.

- **Marijuana** is most often referring to the plant or its preparation to be used as a psychoactive drug. The word itself is first attributed to Mexican Spanish, but the true source is still unidentified. The attribution of the term from a Mexican woman named María Juana is likely a folk tale, though it did give rise to the colloquial term for marijuana, 'Mary Jane'.

- **Cannabis** stems from the Latin genus for the entire family of cannabis plants. The terms marijuana and hemp both apply to various cannabis plants. The two most commonly used varieties of cannabis are cannabis sativa and cannabis indica. These two varieties may also be written in shorthand as C. sativa and C. indica.

- **Hemp** is a term for many of the tall-growing varieties of the cannabis plant that are used for the production of hemp products such as rope, textiles, oil, food, wax, and fuel. The varieties of cannabis that are used for the production of hemp products exhibit a lower concentration of the psychoactive chemical compound THC (tetrahydrocannabinol) than the low-growing varieties used for recreational or medicinal drug use.

# A History of Hemp

The cannabis plant is one of the oldest crops known to be cultivated by mankind. For thousands of years we have grown the plant as hemp, utilizing the durable fibers that make up the plant's stalk. Archaeological evidence shows that humanity has relied on cannabis in this capacity since as early as 6000 BC, using the processed fibers to produce rope and cloth.

Hemp has also been a traditional part of the historical and ceremonial practices of many cultures. It has been used as incense, consumed as a meditative aid, smoked recreationally for pleasure, or worn as part of ceremonial clothing. Evidence dating back to 4000 BC demonstrates the cultivation and use of hemp in China. Hundreds of years before the birth of Christ, holy texts of the peoples of India describe the use and effects of hemp's intoxicating resin. Over 5000 years ago Chinese herbalists prescribed hemp medicinally for a wide variety of health concerns. In Africa hemp was—and still is—used to treat the effects of venomous snake bites and to aid in childbirth. Hemp has been used and grown on every continent that people inhabit for thousands of years.

### In The United States

Hemp production is no stranger to North America, though considering the way that it is currently restricted through legislation, one may think otherwise. Historians aren't sure how hemp made its way to the new world. It is thought that it may have been brought by Chinese explorers or drifting shipwrecks. Another theory is that the seeds may have been brought across the Bering Strait by migrating birds.

The British Crown decreed that American farmers grow hemp in the Virginia colony for export back to Great Britain throughout the 1700s. At that point, hemp was being used for a much larger variety of applications than just rope and cloth. While many of the seafaring ships that were produced by the British Navy—and later the first Navy of the United States—used hemp rope for their riggings, hemp was being processed into paper as well. In the spirit of rebellion, one of the first protests against the Crown was the resistance to the mandatory growth of hemp by American farmers, an ironic twist considering the restriction placed on it today. Additionally, the Declaration of Independence was drafted on hemp paper, making the cannabis plant as American as fireworks and apple pie. Many of our nation's founding fathers grew hemp and were vocal advocates of the versatile plant's multitude of uses. George Washington, Thomas Jefferson, and John Adams were all hemp farmers as well as patriots and statesmen. In the mid 1800s Abraham Lincoln used a fuel processed from hemp seed oil to light household lamps. The United States Department of Agriculture (USDA) showed in 1916 that, per acre, hemp produced up to four times more paper than did trees[1].

What seemed like an unending upward trend for hemp crashed to a halt in 1957. The US government included hemp varieties of cannabis in the same illegal category as those used for the psychoactive drug, and at the same time incentivized the production and development of alternative industries. Hemp farmers were bankrupted in a short time, and legal hemp production in the US all but ceased. It has only been recently, within the last decade and a half, that hemp has been reintroduced into the United States via imports from foreign growers. Often the imports are of processed hemp products such as seeds, oils, and fibers and not of the plants themselves.

## The Next Step

Now that you know a little more about the storied history of the cannabis plant you can start learning how to grow your own. Whether you are a recreational user attempting to keep yourself supplied with a little extra for your friends or an ambitious grower who is attempting to start an entire farm or extensive growing operation, the basics included here will get you off to a great start.

[1] USDA Bulletin No. 404 1916

# CHAPTER TWO
## KNOWING YOUR GROWING

The good news for aspiring cannabis growers such as yourself is that marijuana is an easy plant to grow and cultivate. It is simple to maintain and is strong enough to survive a wide range of growing conditions. This is an asset to growers who farm the plant outside of the bounds of the law, and many have become adept at growing in the most unlikely of conditions—more on concealment later.

Cannabis is a type of plant known as an annual. Annual plants complete their life cycle in a single growing season. This means that in the wild they die off for the duration of the winter and sprout anew in the spring. The opposite of annuals are perennials or plants that grow for three or more seasons. While perennials are a good choice for decorative gardens and landscaping, we won't get into them here. For the grower this means that the annual cannabis can be planted from seed, grown to maturity, and harvested in one season. Due to selective breeding and careful cultivation, the hardy cannabis plant can be kept alive for much longer than a single growing season and can be harvested for many years if developed properly. All of this is good news for cannabis farmers both for profit and for recreation. After all, low costs mean higher profits. Cannabis is further identified as a flowering annual, and its flowers are the plant's method of reproduction. The flowers—often referred to as buds—are also the source of the highest concentration of the psychoactive chemical, THC, and this portion of the plant is harvested for recreational or medicinal use. While there are numerous cannabinoids present in the plant's buds, THC is the most potent and is often the most sought after. Who knows though, the pace of modern research could reveal other uses for one of the many other chemical compounds found in the cannabis bud.

The basic needs of a cannabis plant are light, water, an environmental temperature kept between 70-80°F (21-27°C), and sufficient nutrients.

When these needs are met, the cannabis plant can grow for many years and provide the grower with yield after yield of potent buds. Though there are a variety of strains for a grower to choose from, there are two primary types of cannabis: cannabis sativa and cannabis indica. The chemical properties of each of the plants is beyond the depth of this book, but we'll look at how the two varieties generally differ.

Cannabis indica grows as a short and stout plant. Users claim that the effects of this plant are often more calming and relaxing. Cannabis sativa grows as a taller and narrower plant. It is often said to produce more of an energizing and stimulating effect. It is important to note that an incredible number of variables can affect the way that users experience the effects of consuming marijuana. The effects often vary from user to user, but some variables include personality, frame of mind, intent, experience (or naiveté), method of consumption, and tolerance. We'll explore strain selection in a later chapter, but keep the desired effects in mind when we revisit the topic of strain selection.

## Stages of Growth

The cannabis plant grows in three major stages: germination, vegetative growth, and flowering. From these major stages growth can be further broken down into six more basic stages: seed, initial growth, seedling stage, initial vegetative growth, pre-flowering stage, and flowering. See Table 1 for a summary of each of the growth stages along with their estimated durations.

| STAGES OF GROWTH FOR THE CANNABIS PLANT |||
| --- | --- | --- |
| **Germination** |||
| Seed | Initial Growth (1 - 21 Days) | Seedling Stage (7 - 21 Days) |
| Seed is planted in soil | - Roots are formed<br>- Two initial leaves | - Plant is established<br>- New leaves are established |
| **Vegetative Growth** |||
| Initial Vegetative Growth (2 - 3 Months) | Pre-Flowering Stage (Up to 14 Days) ||
| - The plant grows taller<br>- The stem starts to become a stalk | - Development of calyx<br>- Growth slows ||

| Flowering |
|---|
| (4-6 Weeks) |
| - Develops flowers |
| - Sex of plant can be identified |

*Table 1*

## Germination

The first basic stage of growth is called germination. In this stage the seed is in the initial stages of growth. The seed's outer layer breaks open and a root erupts from within the seed. The germinating seed continues to form roots that push downward into the soil or growing medium. Above the soil two of the initial leaves begin forming and growing upward. We all know the basics: roots hold the plant in place and gather nutrients and water from the soil while leaves collect sunlight and perform respiration. The germination stage of a cannabis plant lasts anywhere from a day to three weeks.

## Seedling

Now that the initial leaves are growing upward, they can absorb light and provide energy to the developing plant. This powers the growth of additional leaves and the development of a discernible central stem from which the leaves branch out. The seedling stage of development may only last one week, or it may last as long as three weeks. After the seedling period the plant has developed four to eight leaves.

## Vegetative Growth

Now that the plant is established as a seedling, it begins the stage known as vegetative growth. The plant continues to grow taller, and the thin stem becomes bulkier to form a stronger stalk. Not only are more leaves growing, but the leaves mature further to demonstrate the characteristic marijuana shape. Vegetative growth occurs over a period of a few months.

## Pre-Flowering

The production of flowers is necessary for the plant's reproduction, but it is also highly consumptive of the plant's energy. With an expanded root system and more leaves to gather light the plant begins to fill out and prepare for flowering. Vertical growth is reduced for this stage as the plant is directing its efforts to the production of flowers instead of reaching toward the light source.

In this stage, the plant begins to show signs of its sex, and in preparation for flowering calyxes, it begins to grow where individual clusters of leaves meet the stalk. In terms of botany, the scientific study of plants, a calyx is the plant structure that eventually becomes a bud that identify the sex of the plant. Calyxes on cannabis plants are small protrusions that mature within the flowering stage. Pre-flowering may occur over a period of up to two weeks.

### Flowering

The flowering stage is the final stage of the plant's growth cycle. This stage lasts from four to sixteen weeks and is the level at which a plant's sex is clearly visible. When the plant produces flowers, the shape of the flower determines its gender. Male plants produce clusters of small balls that become pollen sacks. Female plants produce pistils that resemble fine hairs. Pollination occurs when the male pollen sacks burst and pollen carrying the genetic material of the plant interacts with the pistils on the female buds.

After the flowering phase of the plants, pollinated female buds produce seeds that mature within the fertilized bud for up to another sixteen weeks. When the seeds reach maturity, the pods burst and the seeds drop. In the wild, the seeding process ensures that future generations of the cannabis plants return the following growing season after dying off for the winter. Remember, cannabis plants are annuals.

## Importance of Gender

So what if one plant is male while another is female? What does that have to do with anything? The answer is everything. It is extremely important to understand how to sex your plants to ensure the most productive crop yields. We already know that cannabis plants are annuals, but they are also dioecious, meaning that plants produce either male or female flowers but not usually both. The opposite of dioecious is monoecious, or a plant that has both male and female flowers present on a single plant. There are rare instances of cannabis plants that are hermaphrodites with male and female organs present in each bud or both sexes of flower present on a single plant.

While it may sound advantageous to have cannabis plants that will self-pollinate or a room full of both sexes that will propagate much faster, the opposite is more desirable for growers, especially for-profit growers. If the plants are being cultivated for recreational or medicinal purposes, then the concentration of THC present in the buds will be the chief indicator of quality. When male and female plants are mixed within a generation, or even within an entire crop, this can have a profound and negative effect on the levels of THC that are found within the harvested buds.

The pollination or reproduction process for the cannabis plant involves male pollen sacks bursting and the pollen transferring genetic material to the pistils of the female bud. Once the female bud is fertilized, it begins to produce a seed. The problem with this scenario is that once a female plant is pollinated, it directs all of its energy to the development of the seeds. Don't blame her, you would do the same thing. Once the seed is growing, the levels of THC produced in the bud drop off and the result is a much less potent product. As a grower you want to cultivate a high-quality, potent product no matter what its end use will be. A quality product can command a higher price, and the process of separating your male and female plants is relatively easy.

## Sexing Your Plants

A farmer who raises livestock will tell you that if you want to know the sex of one of your sheep, just lift the tail. Not so easy with plants that have neither tails nor what we would recognize as genitals. If you are starting a crop with a random batch of seeds, it is likely that you will have an approximate split of 50/50 male and female plants. There is no way to tell which sex the plant will be as a seed; the first indicators only come once the plant reaches the pre-flowering stage. It's easiest to identify the gender of a plant once it has completed the flowering stage, but by then it may be too late. The males may have successfully pollinated the female plants and an entire crop, or at least a portion of it, may be ruined.

Remember calyxes? They'll begin to form in the fourth or fifth week of growth. To refresh your memory on the entire growth cycle, refer back to Table 1. Close inspection of these soon-to-be buds is key to successfully separating your plants early enough to avoid accidental pollination. This is a key point to the successful growth of marijuana; we will revisit these visual indicators throughout to book.

| GENDER DIFFERENTIATION SIGNS ||
|---|---|
| **Male** | **Female** |
| Calyx will produce signs of pollen sacks | Calyx will produce signs of pistils |
| Pollen sacks resemble small balls | Pistils resemble fine hairs |

*Table 2*

You may not want to wait until your plants reach the flowering stage before identifying their sex. Remember how low costs can contribute to high profits? Cultivating an entire crop of plants requires time, effort, and resources. It's not prudent business to spend time and money growing a crop of plants only to discard half of them. Effectively this could double the cost of growing your plants before you can correctly sex them, so identify the gender as soon as possible to save time and money. The following are quick sex identification techniques. Keep each of these in mind when sexing your plants as they are not all guaranteed to be 100% accurate—where one may not work, another could save you time and money by weeding out males quickly and leaving the THC-rich females unfertilized. The exception to this rule is cloning, which is a little more involved and highly reliable.

### Rate of Maturation

A basic indicator can be the time it takes each plant to show the signs of its sex, otherwise known as maturing. Females will often take longer to show signs of budding, so if a bunch are showing signs of being male and the others are taking their time, this in itself can be an indicator of which plants are female.

### Buying All-Female Seeds (Feminized)

What if you could avoid growing males in the first place? Growing from a batch of only female seeds will ensure that you spend less time worrying about the gender of your plants. Methods of producing all female crops for seed sale are outside the scope of this book, but some sellers may be able to provide pre-sorted seed batches for larger growing operations. Be aware: even in a female-only batch of seeds, a male or several may pop up, so still pay attention. Feminized seeds are coved in greater depth in Chapter 6 of this book.

### Size

Other early sexing techniques include observing the size of your plants as they grow. Male plants tend to be larger than the female plants that were planted at the same time and have grown under the same conditions. While this is generally considered to be quite effective at distinguishing sex, it is not 100% guaranteed that female plants will be smaller, so keep that in mind.

## Cloning

Cloning plants is the only true way to guarantee their sex. A clone will have the exact same DNA as its host or parent, and therefore the exact same sex. While this is the most involved and time consuming method of sexing your cannabis plants, it is also highly accurate and the only reliable way to ensure correct gender identification. The cloning process involves taking a small cutting of the parent plant. The cutting is placed in soil and allowed to grow on its own for a few days before it is forced into early flowering through a process of 12 hours of darkness and 12 hours of light. This process is done apart from the host plants as they should be allowed to grow normally without forced flowering. Once the clones begin to flower, their sex can be determined. Remember to devise a system to ensure that the clone is matched to the parent from which it was separated, otherwise cloning is a waste of time.

## Sprout Location During Germination

While this is perhaps the least scientific of all the early sexing methods, many growers claim that they can identify the sex of their plants with 90% accuracy just by observing where the sprout emerged from the seed. It is the experience of many growers that top or bottom sprouts result in female plants, whereas side sprouts result in male plants. Don't start throwing away all of your side-sprouting plants just yet; keep a log of your plants that sprout sideways and see if they end up being males. If you are seeing consistent numbers, consider using this method of early sexing. Some growers trust it to the extent that all side-sprouting seeds get discarded immediately while others are more suspicious of the consistency of the results.

## Hermaphrodites

Keep an eye out for plants that exhibit both pollen sacks and pistils. If these occur, try pruning off the male calyxes in an effort to train the plant to only develop female flowers. A hermaphroditic plant could potentially self-pollinate or pollinate other nearby plants due to the presence of the male flower and its pollen sacks.

Growers who are producing marijuana for recreational or medicinal purposes will be harvesting from the female plants, the ones that produce the highest levels of THC. The earlier males can be detected, the lower the chances they have of pollinating the females and the fewer resources are consumed raising useless plants. Once the males are separated they can be destroyed, or they can be reserved to pollinate more females in the interest of breeding your next crop.

Try different methods and evaluate the results for yourself. Cloning is the most reliable method, but it takes time and resources. We all know that time is money, and money is money, so if you can determine a method that's quicker and gives you a level of reliability that you're comfortable with, that's the best choice for you.

# CHAPTER THREE
## FINDING A SUITABLE LOCATION

Of course knowing how to grow your cannabis plants won't do you much good if you have nowhere to grow them. Finding a suitable location is one of the earliest—and most critical—decisions that a grower must make. This means putting a lot of thought into your selection and taking many factors into account. This chapter focuses on how much space your plants need indoors and outdoors, and how many plants you should grow to reach your goal.

When selecting a location for your growing operation, knowing the size (or target size) of your crop is paramount. This applies to growing both outdoors and indoors. If you, as a grower, want to raise a crop of a thousand plants, you should be aware that this simply can't be done in a broom closet. Ask yourself these questions:

### How many plants do I want to grow?

Knowing how many plants you want to grow may be a bit like putting the chicken before the egg, but it is important to have a plan. You know why you're choosing to grow marijuana; whether to build a massive cash-generating farm or to keep a pair of plants in the closet for recreational use on the weekends, and anything in between. If you're planning on expanding or you intend to eventually have a large operation, keep scalability in mind.

Every new project should consider scalability, and in the world of business it is a good practice to assume that your competition is doing everything right. If your competition is considering scalability, then you should be, too. Scalability is the operation's ability to grow effectively as demand and resources increase. Ignoring how to scale your operation when it is still at the foundational level can cause some serious problems down the road. Remember, Rome wasn't built in a day. It must have taken at least two or three. Kidding aside, if you're not planning for expansion, you could end up in over your head. For the prospective new grower this means selecting a location that is much larger than your new operation currently warrants - a space that you can grow into, so to speak. If you select a location that will only support 10 plants because that's how many you can raise by yourself, that's fine if you want to have 10 plants. But if you have aspirations of expansion, finding a larger space early will mean that once you have the capital, you can turn the 10 into 30 or 60 depending on the space you have.

Not everyone has the luxury of a selection of easy-to-conceal and spacious growing locations, and that means that you will have to mold your plans to fit your situation. Such is business, and such is life. As they say in the military, no plan survives contact with the enemy. Don't just ask yourself this first question; read this entire book and come back to the process of location selection taking into account all of the factors and information you have at your disposal when making your decision.

Realistically experts agree that small operations should start with 2-6 plants. Four plants are much easier to care for than sixteen, and when growers are attempting to grow in areas where cultivation is illegal, the lower the number of plants attached to your name, the better off you will be if you are caught. For indoor growers, yields are better when fewer plants rely on each light source, and for outdoor growers there is a minimum spacing between plants to produce ideal results. The monetary incentive to pack plants into a space may be attractive, but overcrowding will reduce yields.

## Using a Target Income

Financial planners will tell you that the way to build a business or a financial plan is to set a goal of income and build your business around that goal. If that's the route you want to take, my advice is start with a reasonable goal. Assess the costs of setting up your operation and what your potential profit will be. A key metric in this decision is knowing how much product you can expect to get from a single plant. As you might expect, the answer varies, but we'll take a look at some industry standards.

For those electing to grow outdoors, growers can expect yields of up to 500 grams (17.5oz) per plant. That is under ideal conditions, mind you, and under the care of an experienced grower. Remember, no plan survives contact with the enemy; you will only get 500 grams per plant if no one steals any of your crop, your plants don't suffer from disease or pests, and you can provide them with adequate sunlight, water, and nutrients. To bring yourself closer to the 500-gram yield, we'll discuss growing methods and strategies in later chapters.

For those electing to grow indoors, yields can vary wildly with the setup of your growing operation. Indoors, lights are key; an experienced grower can expect to produce nearly a gram of product per watt of light. Again, this is under ideal conditions and if everything goes according to plan. Too many plants relying on a single light will result in the same overcrowding issues that outdoor growers will face.

Can you expect 400 grams of product from a 400-watt light? Six hundred grams from a 600-watt light? The answer: more or less, yes. The yield per plant varies significantly when growing indoors, and the grams per watts ratio is an average for the entire operation. If you grow four plants under a 600-watt light, you can expect to see about 150 grams per plant, meaning that your production overall will be about 600 grams. If you pack the plants under that same light, the yield per plant will be lower, but the total will still be around 600 grams.

How do you translate this information into a rough idea of how many plants you want to grow? If you have a target yield and you know how much money you can produce from a single gram, then you are well on your way. Know what selling structure you will be using: wholesale or end user. If you sell to people who in turn sell, they need to be able to make a profit from individual sales, hence wholesale pricing. You can sell large amounts of product and walk away with cash in hand. If you elect to sell directly to consumers, there are more expenses incurred and you will need many more transactions to turn your harvest into cash.

Once you have determined how much money you can make per gram and how many grams you need to hit your target income, subtract the costs of equipment and growing, and the result is a basic projection of income. Generally speaking it makes more sense to produce more buds per plant on fewer plants because this saves on costs, but it also means that your entire crop is at risk in the event of disease, theft, power outages, or just about any other catastrophe you can dream up.

So if you only need to sell 2,000 grams to hit your target income, think about your yield per plant. That's four outdoor plants for an expert, but I'd recommend doing it with six or seven until you get the hang of things. I doubt you'll complain if there's a surplus. Or, that's 2,000 watts of light indoors. Two thousand watts can be organized as two 1,000-watt lights, ten 200-watt lights, and so on. Remember, growers can effectively grow many plants under one light, but as the number of plants increases, the yield per plant decreases. Fewer plants mean higher yield per plant and lower costs but higher risk to the crop. If you have four plants and lose one, that means roughly 25% of your harvest is gone. More plants mean higher costs and more work. There's less yield per plant, but the loss of a single plant is a lower hit to your harvest. The choice is yours.

## How tall will they be?

Outdoor growers have less to worry about in this arena, but as everyone knows, plants grow upwards. When considering the space that a crop will occupy, you must also consider the vertical space into which it will grow. The strains you select have a known average height, so take this into consideration. The ballpark figure is that cannabis plants can vary between heights of only a few feet tall to roughly ten feet in height. The largest factor that impacts this is whether your strain is a sativa or indica. Remember that the former is the taller of the two and the latter is shorter. Sure, this means don't grow marijuana in the low ceilinged rooms of your house, but also consider this: are you growing on tables or on the floor? Tables are handy for care when the plants are still small, but if you have a basement operation, the standard height for a basement ceiling is a minimum of 7 feet. If your plants are on 4-foot tables and grow more than 3 feet tall, you'll be left with squashed and stunted plants. Have a plan in place if you'll need to transition from tables to floor.

The height of the plants is not only a concern for growing space, but it can potentially be a concealment concern as well. When planning the concealment of your growing operation, consider the final height of the plant. If your concealment will hide a 10-foot tall cannabis plant, it will hide a 3-foot tall one; the inverse is not necessarily true.

## How much space will the plants occupy?

Now you know how many plants you want and how tall they'll be. How much space do you need? Think three dimensionally because your plants could potentially be 10 feet tall if you plan on growing a sativa strain. The term that describes how closely your cannabis plants are packed is planting density. The density of your plants can vary based on strain, so as a result a lot of general guidelines are offered. Experts agree that effective density can be a stumbling point for new operations because it is a balance between several different metrics.

When growing indoors, growers want to get the most out of each watt of light and square inch of space. Electricity isn't free, and the high-intensity lights that growers use are energy hogs. Having a high planting density means that you will get the most bang for your buck so to speak when it comes to not wasting precious kilowatt hours (the way your energy company measures your consumption of electricity).

Having too high of a planting density means that while your harvest weight will come in at target, the actual quality of the buds will be lower. Whether you personally use marijuana or not, it should be fairly obvious that sub-standard marijuana is bad for business. What works best for one strain won't necessarily work as well for another, but a good rule of thumb is to keep six strong branches per linear square foot of growing light. The light output of different lamps will vary, but the manufacturer will always be able to provide you with an effective discharge range if it's not printed on the packaging. This guideline is for the production of the highest quality product, and small compromises in quality can be made in the interest of increased quantity and cost effectiveness. The bottom line is that no matter how many general guidelines you read, there is simply too much variation from strain to strain for any single guideline to be accurate for your operation. Your seed dealer will be able to recommend an approximate planting density for your strain

## Can I effectively care for this many plants?

As you become a more skilled and experienced grower, the level of care that you provide for your plants will increase. When starting out, your care and maintenance duties will be fairly straightforward. When you become more adept, the job of growing can become much more labor intensive. There is no formula to determine how much time of your day will be devoted to caring for your cannabis plants. It should be fairly obvious that caring for fewer plants is much easier than caring for a bunch of plants that you maintain, all the way up to an entire field of marijuana with an entire team of growers.

As you read on and learn more about the specifics of plant care, think about each of these activities, its level of difficulty, and the amount of time that you have available (or will make available) to develop your growing operation. After all, the best decision is an informed one.

## Should I grow indoors or outdoors?

The answer to this question is up to you, but before you decide there are some aspects of cannabis growing that you should consider. The first and foremost piece of information that you should consider is the level of concealment you wish to implement. It's simply easier to hide your growing operation indoors and away from prying eyes. That being said, in some instances growers prefer to grow outdoors in an effort to gain some measure of plausible deniability. As I have said, I am no lawyer, but I wouldn't bet on that keeping someone out of jail. Another benefit to growing indoors is the level of control it affords the grower. When growing outdoors your plants are... outdoors. They're subject to the weather, the elements, soil conditions, insects, and you name it. Getting a more consistent harvest and plentiful harvest out of outdoor crops can be harder—especially for the novice grower—and it can expose your growing operation to risk in the form of loss or discovery.

When working with cannabis, indoors growers don't have to worry about a cloudy day affecting their harvests, acid rain stunting their plants, or depleted soil starving their crops. All of these examples are of course worst case scenarios, but you get the idea. The reality is that cannabis is a versatile and hardy plant, so it is not as susceptible to environmental stressors as other plants are, but your growing operation is a business. You're not trying to raise plants that simply survive, you want productive plants that will thrive.

The specifics of running an indoor growing operation are covered later in this book. Again, keep all of these questions in mind throughout the next several chapters. Answering these questions will ensure that once your operation is functional, you can hit the ground running and avoid costly and time consuming mistakes.

# CHAPTER FOUR
## CONCEALMENT DO'S & DO NOT'S

Maintaining discretion and secrecy around your growing operation is essential, even though you probably want to show all of your friends your budding operation. When concealment and marijuana are in the same sentence, the first thought is concealment from law enforcement, and that's a very real concern for many growers. Remember, the author is not a lawyer and cannot give legal advice. Understand the laws and regulations in your area concerning the growth, sale, and consumption of marijuana.

Another aspect of growing concealment that's often overlooked is concealment for security purposes. Even in areas where growing marijuana is legal, there are many reasons to conceal your growing operation. The first is outright theft. A season's harvest of quality buds is valuable. An open field of mature females would be easy pickings for a group of motivated marijuana thieves; an effective team could strip a field in a single night under the cover of darkness. There goes your crop, and there goes your cash and hard work. Even if no one is stealing your buds, as a grower you probably don't want strangers loitering in your grow operation. At the very least they won't be as careful with your plants as you would be, and they could cause minor damage to some plants. Damage to plants is damage to profit. On the other end of the spectrum, they may tell others about the scale, profitability, and location of your operation. While someone who may not himself steal from you isn't an immediate threat to your operation, he could lead others with less honorable intentions right to your doorstep.

Other reasons are less serious but should still be taken into account. Nosy neighbors, attentive landlords, and pesky parents or family members should all be considered when selecting a growing location. In some locations where it is legal to grow marijuana, some townships or cities have regulations about the use of specifically zoned land for the growth and cultivation of crops. Again, concealment plays a role here, but don't say I didn't warn you if a code enforcement officer discovers your crop. In areas where marijuana growth is illegal, you have significantly more to worry about than zoning enforcement violations.

Concealment is not always easy, and in some cases growers have to make tough decisions. If an area is easy to conceal but difficult to reach, and care for the plants is difficult due to the location's remoteness. Should the grower select a more accessible area and sacrifice some measure of concealment? Finding a balance between concealment and a location's conduciveness to growth can sometimes require some tough calls on the part of the grower.

## The first rule of growing marijuana is 'Don't talk about growing marijuana!'

Unfortunately there is not one perfect, one-size-fits-all concealment method. The efforts that individual growers undertake to conceal their operations vary as each grower's circumstance is different. The first thing to remember about successful operational security is to keep your secret exactly that: a secret. This means don't tell anyone. This book is about how to successfully grow marijuana, not about how to pick friends. No one is telling you not to trust your friends, but the fewer people you tell about your operation, the fewer people know. The fewer people know, the safer your secret is. There are many ways for growers to slip up; if you keep your operation a secret, then that's one less thing you have to worry about.

### Low-Traffic Areas

Think about how much foot traffic or interest your prospective growing operation's location generates. Many businesses are looking for exposure and visibility, but marijuana growers are looking for the opposite. Think about the city you live in. There are literally thousands of places that could right now be home to an opportunistic grower's operation. A back office over a bar, a rundown shed that is set back from the street, a disused basement... the possibilities are virtually endless.

Closets and basements are good choices because they're often low-traffic, out of the way, and will mask the details of your growing operation. A closet in a guest bedroom is a better idea than a closet in a master bedroom. Think about how often the area is disturbed and how often residents of the house will go to that area. Obviously if you live alone, this isn't an issue, but don't forget service personnel. Will the cable guy have to wade through a sea of cannabis plants to get to your box? Will a maid be tidying up in that room or closet?

Basements are often the best choice for indoor growers because they mask all of the telltale signs. Marijuana crops have a pungent and distinctive odor. Inches thick concrete walls that are mostly buried will help keep the aroma contained. Remember to tape and black out any casement windows. In addition to disappointing any snoops, it will prevent any of the bright growing light from escaping. Bright lights in a basement at odd hours are a bit of a red flag for a growing operation.

When growing outdoors, think about the property. Do you own the property? Some growers squat on someone else's land to cultivate their cannabis assuming that if caught, they can pass the blame onto the property owner. This isn't a great idea for a number of reasons, the chief among them being that, everything else aside, in order to get to your plants you have to trespass on someone else's property. They can claim that the plants are their own or destroy them at will, eliminating your crop and your investment in one fell swoop. Outdoor growing operations shouldn't be visible from high traffic areas such as pedestrian walkways or streets. Ideally they should be disguised as something else, something that is functional enough to warrant some visits but dissuade people from nosing around. One huge advantage to outdoor growing is the cost savings it provides. The sun will rise every day and provide you with free and abundant light; the same can't be said for indoor grow lamps. This also helps keep your operation concealed as law enforcement has been known to identify grow houses when energy consumption spikes from a baseline normal consumption to much higher usage when energy hungry grow lamps are introduced.

**Consider this**: if you have an abandoned house with no utilities hooked up sitting on a plot of land, and passersby see your car there frequently, it raises suspicion. The same thing goes for remote areas of shared property. Low traffic works the other way too; in addition to helping hide your growing operation by limiting the number of people that may come across your plants, it also arouses suspicion based on the fact that people wouldn't frequent that area.

## Maintain a Low Profile

This is the culmination of the previous two tips. Don't tell anyone, and keep the traffic surrounding your growing operation in mind. Don't get into a routine. Routines are predictable, and if someone wants to know what you're up to badly enough, she can use predictability to stay one step ahead of you.

If you keep a growing operation in the woods and your 'cover story' is that you're a hiker, don't forget to wear hiking boots and carry some water and snacks. No one is going to believe you're on a hike wearing flip flops and wearing a 'Legalize It' shirt. The same goes for supplies for your grow room or outdoor grow. Don't carry light after light into the house. Use cardboard boxes that obscure the contents or black contractor bags to move suspicious supplies. Part of maintaining a low profile means not attracting attention—that much should be clear. This means that you should avoid getting into scrapes with the law. If you are a frequent recipient of parking tickets or noise violations, it is in your best interest to tone it down. When a party gets out of hand or music is blasting, it's your neighbors who are calling the cops. No music, no outrageous parties, and the cops are nowhere in sight. If your neighbors believe that you have settled down, they'll butt out of your business as well.

There's no reason you can't have fun, it is your house after all, but keep in mind what is at stake, especially for those growers who live in areas where the cultivating cannabis is still illegal.

# CHAPTER FIVE
## FILLING YOUR TOOL BELT

Any professional will tell you that without the right tools for the job you won't get very far. The same goes for the cultivation of cannabis. So far we've talked about a lot of the theoretical aspects of raising your own crop of marijuana: understanding the plants, where to grow them, how to plan your operation, and how to keep it a secret. Now it's time to roll up our sleeves and get to the good stuff.

The extent of equipment that is necessary varies from operation to operation based on the sophistication of each, the level of concealment, the size, and the intent. Some growers want to produce crops of high-quality buds exclusively for sale, while others want a few dedicated plants for personal use. The sophistication and effectiveness of your growing operation depends on the equipment and know-how that you have. As you read this section, keep your equipment needs in mind and refer back to them when considering a space for growing. Ready access to power is one of the biggest concerns when growing indoors—something many closets sorely lack.

## Lighting

Unless you're growing entirely outdoors, you will need an artificial light source. Even if you're growing outdoors it may be prudent to start plants indoors or differentiate the sexes before transporting them to the outdoor location. Before we get in depth about the lighting you'll need, let's take a second to talk about safety.

### GFCI: SAFETY FIRST

You have been on this planet long enough to know that electricity and water don't mix. Always keep safety in the front of your mind when planning and setting up your grow operation. Plants can be replaced just like other property; electrical shorts can kill. If you don't already have access to Ground Fault Circuit Interrupt (GFCI) outlets, take the time to look into them. For those of you that may not know, GFCI outlets snap the circuit closed in the event of a short to prevent electric shorts, sparks, and fires. GFGI outlets can be easily identified by the presence of two buttons labeled 'test' and 'reset'. These buttons are integral to the use of the outlet, and if a power strip or multi-plug is used to power multiple electric appliances, keep in mind that they all will trip the GFCI in the event of a short. In many cases it is an easy, DIY job to upgrade normal household outlets to GFCI protected outlets. Installation is easy and requires some inexpensive hardware and just a little technical know-how. The process doesn't require any electrical expertise, but if you are unsure or uncomfortable doing any kind of work around the house, consult an electrician.

If you already have access to all the GFCI outlets that you need, then you can skip the following installation instructions.

**Remember to always turn off any circuit that you will be working on prior to beginning installation. Always test outlets to verify that they are not part of a live circuit.**

1. Remove the outlet cover plate with a screwdriver and remove the outlet from the wall box.

2. Loosen the wire terminal screws on either side of the outlet. There will be a white wire, a black wire, and a bare copper or green wire. Each of these wires will be affixed with the following screws: white wire, silver screw; black wire, copper screw; and bare wire, green screw. If your outlet does not have all three of these wires, it is GFCI incompatible. To upgrade a circuit without all three wires will require the assistance of a professional electrician.

3. Once the old outlet is removed, follow the installation instructions that accompany your new GFCI outlets. Keep in mind that you may need to use needle nose pliers to reconnect the wire terminals to the new outlet screws.

GCFI outlets run approximately $20 at many hardware stores.

## For Bigger Operations

Bigger growing operations require a different approach. High-output lighting can have a massive draw on a circuit, especially if several lamps are strung together from the same circuit. A GFCI outlet will protect you and your property from shorts, but it won't change the max load of the circuit. While it may be tempting to daisy chain power strips to produce a line of outlets, too many lamps can and will overload your circuit. Consider a dedicated circuit breaker and possibly a sub panel to provide adequate power. Not only will a dedicated circuit be more convenient, but it will also allow you to optimize your power and ensure that your life-giving lamps are operating at capacity. If you want to get fancy, you can install timers and smart meters into the circuit to ensure that every spare electron is saved and each penny is accounted for.

The installation of new circuits, panels, and extensive wiring work is definitely in the realm of the professional electrician. It is not recommended that you attempt a rewiring of your basement as a weekend DIY project. Consider using a dedicated circuit if your lighting needs exceed 1000 watts. The instructions to rewire a house are well beyond the scope of this book.

How do you know if your needs will exceed 1000 watts?

## Determining Your Lighting Needs

The light needs of your cannabis plants are broken into two broad categories based on the maturity of the plant; whether each plant is in a vegetative stage or flowering stage determines the best light cycle.

During vegetative growth your cannabis plants will need a standard daylight cycle: 18 hours on and 6 hours off. There are some growers that will advocate a 24-hour on cycle as it stimulates quicker growth, but this may vary from strain to strain. During the flowering stage the light cycle will shift to 12 hours on and 12 hours off. Remember that for many strains the flowering stage begins after 40 or so days.

This type of behavior classifies cannabis as a 'photoperiod' plant, or a plant whose flowering is determined by the amount of darkness (duration of day) it receives. The exception to this type of behavior comes with auto-flowering strains, or strains that produce buds regardless of the light cycle used. Auto-flowering strains are discussed in the next chapter.

The amount of light the plants should receive isn't the only factor that changes with the growth of the plant. The kind of light that produces the best results changes, too. Any kind of fluorescent bulb is best for the vegetative stage of growth, as the kind of light they produce is rich in wavelengths that fall into the blue spectrum. Of course most of the light looks the same to our eyes, but the blue spectra light emitted by fluorescent lights promotes efficient growth for the vegetative stage.

## Fluorescent vs. Metal Halide

So if fluorescent lights work, then metal halide lights should work better, right? Yes, and no. Fluorescent lights ionize mercury vapor in a glass tube, which excites electrons in the gas to emit photons at UV frequencies. UV frequencies are considered 'blue' light that is made up of shorter wavelengths and is invisible to the naked eye. The invisible UV light is converted into regular, visible light through a phosphor coating on the inside of the glass tube.

Metal halide lights, on the other hand, are electric lamps that produce light by an electric arc through a gaseous mixture of vaporized mercury and metal halides. Halides are compounds of metals mixed with bromine or iodine. It is a type of high-intensity discharge (HID) gas discharge lamp, and while metal halide lamps produce an abundance of blue light, they are much less discreet and much more expensive to run.

HID lights require much more electricity to operate. More electricity means higher costs and higher chances of detection by law enforcement for growers who cannot grow cannabis legally. Most growers find that fluorescents do the job well and at a reasonable cost, though professional growers will make a case for the HID style lights for an absolutely top shelf final product.

## The Flowering Stage

We know that blue-producing fluorescents are essential for the vegetative stages, but as the cannabis crop enters its maturity and the flowering stage, the light spectrum needs of the plants will change again. The flowering stage requires orange, yellow, and red spectra light, or the type of light emitted from high-pressure sodium (HPS) bulbs.

Sodium vapor lamps, like other types of gas discharge lamps, use electricity to excite the electrons of a sealed tube of sodium vapor. HPS lamps in particular require a ballast or a currency regulator. Without a ballast component, a gas discharge lamp draws increasing amounts of energy from a high voltage source. This is necessary to start the lamp and produce light, but if left unchecked it results in 100% of the bulbs breaking. Ballast components can draw large amounts of energy and produce large amounts of heat. This is somewhat mitigated by the fact that the plant's light cycle is changed to 12 hours on and 12 hours off.

For the savvy grower, Light Emitting Diode (LED) options are available that are specific to various growing spectra needs. LEDs consume considerably less electricity, produce much less heat, and have much higher lifespans than other types of bulbs, but they are considerably more expensive up front.

| LIGHTING NEEDS DURING STAGES OF GROWTH ||
|---|---|
| **Vegetative Growth** | **Flowering Stage** |
| Normal light cycle: 18 hours on, 6 hours off | 12 hour light cycle: 12 hours on, 12 hours off |
| Fluorescent or metal halide lamps: plants require light in the blue spectrum | HPS or high pressure sodium lamps: orange, yellow, and red spectrum |

*Table 3*

The color and intensity of light is based on its particular wavelength. Wavelength is measured is nanometers (nm). The smaller the nanometer measurement, the shorter the wave and the further down the left side of the scale it is. UV light and other kinds of blue light have a shorter wavelength. The larger the nanometer measurement, the larger the wavelength, and the closer toward the other end of the spectrum it becomes. Red light and infrared light are on this side of the wavelength spectrum. For the growth cycle of cannabis, a range of approximately 420nm (vegetative growth) to 750nm (flowering) is appropriate. Don't stress counting the nanometers in your light. If you provide your cannabis with the 'wrong' light at each phase, it will still produce buds. For the absolute best results, highest THC content, and potent effectiveness, giving your plants exactly what they need will improve the overall quality of your harvest.

It is important to note that standard household incandescent lights will not work as growing lights. Incandescent lights, or standard bulbs that produce light by heating a wire filament to the point that it emits light, will not produce light with the right wavelength for cannabis (or other plants for that matter). That being said, standard CFL lights or compact fluorescent lights will have some effectiveness. They will not be as productive or efficient as dedicated grow lights, but CFL lights (the curly kind that fit your household lamp socket) will provide some light in the blue range of the spectrum. Don't expect high quality dense buds from CFL light alone, however.

## How Much of Each Kind of Light?

We know cannabis needs different kinds of light based on its maturity, but that still leaves the question of how many lights will be needed for various sizes of growing operations. Because of the varying nature of different cannabis strains, there isn't really a single answer to this question, but there are a number of guidelines that can put novice growers on the right track. Because light is the primary source of energy for your cannabis plants, the quality, quantity, and length of the light cycle will change the output of your crop dramatically. Many growers experiment with planting density and lighting configurations to determine the best way to squeeze every last gram out of their crops. When trying new configurations record everything. Think of your layouts and methods as recipes. If you want the same product again in the future, follow the recipe. If you want to avoid a poor crop again, make sure you do something different. As with many aspects of life, knowing what not to do is just as important as knowing the right way to do things.

As we touched on earlier with the location selection planning, more light does equal more buds. The problem is that too much of a good thing can still be too much. There is a point at which your plants can start receiving too much light. If your plants begin to exhibit yellowing, spotting, or a burned look—especially without the presence of high heat—then they are receiving too much light. Yellowing and spotting from light burn will start at the edges of the leaves and may persist even if the inside portion of the leaves stay green.

To mitigate the effects of light over-saturation, turn down the lights. Some growers have constructed shades for high power lights that can be removed or installed as needed, and some manufacturers will provide ready-made shades for this purpose. Shades are only recommended if the light cannot be turned down. Consider this: if you outfit a light with a shade, the light will not put out any less light; the shade will only protect the plants by eliminating some of that light. This means that you will be paying for the full light output but effectively 'throwing away' a portion of it. If your lights can be turned down, that is always a more cost effective solution. Remember that when changing the environment of your plants, always try to do so slowly; quick or intense changes can cause unnecessary stress on plants. If you ever do construct a shade for your lights, consider heat output as well. Make sure that in constructing a shade you are not also constructing a fire hazard. Not just a fire hazard, heat can be an issue for your plants too—especially with HID lights. Too much heat can stunt the growth of your cannabis plants, and it can even burn them in the same way that our skin would react to touching a hot plate. If your plants are burning and you have hot lights, the solution is simple: move the lights higher or turn them down.

## Lighting Systems

Let's take a closer look at some of the common lighting systems.

| PROS & CONS OF FLUORESCENT GROW LIGHTS ||
| --- | --- |
| Pros | Cons |
| Inexpensive and easy to set up with very little to no maintenance | Poor efficiency—produces little light considering the amount of electricity used |
| Low heat and electricity—discretion and concealment friendly | Best for the vegetative stage—cannot be used for effective flowering |
| Not too hot or powerful—can be used in an extremely short grow area | Low power yields smaller plants |
| Can be used later for clones or young plants (vegetative stage) even if replaced for primary growing | Plant height can be difficult to control under this type of light |

*Table 4*

### Fluorescent Grow Lights

All things considered, the benefits outweigh the detriments for new growers. There are stories of hobby growers with a few plants in a bucket beside a standard socket with a single CFL in it. While that's not a sustainable business model or a good way to create top shelf product, it does underscore the budget friendly aspects of fluorescent lighting.

This style of system is great for beginners or growers on a budget. It works well for a few plants up to a handful and is often the best option for a short or low-ceilinged grow area.

| PROS & CONS OF HID LIGHTING SYSTEMS ||
|---|---|
| **Pros** | **Cons** |
| Can be used to grow a high number of plants | Cannot be plugged into a standard socket—often these lights come with a hood and ballast component |
| Little to no maintenance | Produce large amounts of heat |
| Available in a wide variety of sizes to fit the needs of different grow spaces | May require venting to alleviate heat buildup |
| Much more effective and efficient than other lightings systems—these styles tend to produce bigger yields | Setup can be difficult, complex and expensive |
| Easy to use, common, tested, tried and true | Though both lights can be used for both vegetative and flowering, growth results are best when metal halide is used for vegetative and HPS for flowering |

*Table 5*

## HID Deluxe Grow Lights

HID lights are the next level for growers who are looking to upgrade and graduate from fluorescent lights. These lights produce better yields and make better business sense. When determining which light is right for you, the first thing you should realize about HID lights is that they will generate a lot of heat. This means that they have to be far away from your plants—between 1.5 and 3 feet. This may not be a problem when your plants are young, but as they get bigger, that distance will still need to be maintained. While it is possible to train your plants to grow shorter, they will still get taller to some extent.

LED growing systems are still relative newcomers to the market, and as such the costs are still very high and the industry is still working out the best application for them. If you're just getting started with a few plants in a closet, then LEDs may not be for you; fluorescent systems are probably a better choice to save yourself a couple of bucks and get your feet wet before making larger investments.

Take a look at the following profiles of some different strengths of metal halide and HPS lights to better understand how they will interact with your crop. For a novice grower high intensity lights are not recommended.

## METAL HALIDE & HPS GROW LAMPS

| Wattage | Coverage Area | Distance from Plants | Average Yields |
|---|---|---|---|
| 150 W | 2 ft² | Min. of 7" from plant tops | 75 - 150 g |
| 250 W | Up to 2.5 ft² | Min. of 9" from plant tops | 125 - 250 g |
| 400 W | Up to 3 ft² | Min. of 12" from plant tops | 200 - 400 g |
| 600 W | Up to 4 ft² | Min. of 16" from plant tops | 300 - 600 g |
| 1000 W | Up to 5 ft² | Min. of 21" from plant tops | 500 - 1000 g |

*Table 6*

## PROS & CONS OF LED LIGHTING SYSTEMS

| Pros | Cons |
|---|---|
| Many growers report improved characteristics with the use of LED grow lights | LEDs are expensive—they have a very high initial cost compared to other systems |
| Highly efficient—LED lights produce high amounts of light and consume the least amount of energy | LED lights for growing are subject to a high number of scams or misleading advertising that is untrue—shop from a reputable dealer |
| Contributes to improved quality throughout the industry as well as the ability for many home-based growers to produce professional quality results and yields | Lower yields per watt than the more powerful HPS lights |
| Many have built-in ventilation components—heat is dispersed away from the plants | Even with included ventilation systems large operations may require additional venting for heat |
| There is no need for a ballast, fan or hood—LEDs plug directly into the wall | LED manufacturers or retailers that cut corners have produced lights that will bleach or burn leaves—damage that can affect crop yield |

*Table 7*

A common theme that comes up when discussing LED lighting systems is the prevalence of scams and spurious retailers. Always buy LED grow lights from a trusted retailer. Always do your homework when it comes to incredible claims that LED manufacturers offer. If something sounds too good to be true, it probably is. Again, participating in the discussion online will allow you to learn about which brands and retailers have done right by the growing community and which have been less than honest.

## Water

Indoor lighting is a complex subject; water is not. Some high-end operations and hydroponic or grow tent operations will use complex irrigation systems to regulate and simplify the watering process. When watering your plants as a beginning grower, tap water will work fine—if you can drink, it so can your plants. If you are an experienced grower you should be using water that is pH corrected to have a value between 5.5 and 6.5. Don't worry about correcting the pH of your water when starting out. Technically the only equipment you'll need for this function is a container to transfer the water from the tap.

### When should I water my cannabis plants?

Whenever the growing medium seems to be losing moisture. A good rule of thumb is to water the soil if it feels dry up to your first knuckle. This is hardly scientific—everybody has differently sized hands—but it is a good place to start. If you want a more precise method, check the soil up to an inch or an inch and a half deep. If the soil at that depth is dry, then it's time to water. If you wait until the soil is completely dry, then your cannabis plants (and your bottom line) may have lost an opportunity to turn water into growing power. Many new growers are impatient and afraid of under-watering; there is such a thing as over-watering. Don't let this be you!

Your plants should never be swimming, or more accurately drowning; pots should be allowed to drain off the excess water from the watering process. Standing water leads to fungal growth and rot below the soil line. If your plant's roots are rotting, then the plant is literally being eaten from the bottom up by hungry and persistent bacteria.

If your plants are drowning, there will not be any visible discoloration, but growth will slow immediately. Otherwise healthy appearing leaves will drop, and if your plant survives, it will most likely become stunted and have poor bud productivity.

When considering the watering needs of your crop, keep the following facts in mind:

- Larger plants require more water.

- Larger containers need to be watered less extensively—the plant inside has a larger root system and the larger amount of soil is better at retaining moisture.

- Hotter air temperatures and dryer air both mean that soil loses more moisture through evaporation and plants need more water.

- Higher humidity in the air causes less evaporation; soil retains more moisture and plants require less water.

- When flowering, your cannabis plants use less water than they did for growth during the vegetative stage.

- Covering the exposed soil with mulch such as wood chips reduces the amount of moisture the soil loses to evaporation.

Your plants will tell you if you have not been giving them the right amount of water. Under-watered plants will wilt and their growth will slow. Left long enough without water, your plants will die.

A side note for those using fertilizer-enriched water; if you are attempting to revive a plant that is suffering the effects of under-watering, do not use fertilizer-enriched water. Use normal tap water until the plant is healthy again and normal growing is resumed. Once the plant has been nursed back to health, the use of fertilizer can resume.

If you want to take the guesswork out of watering, many home and garden centers offer moisture meters that can determine the moisture levels of your soil. These are by no means necessary for the novice grower but can prove useful, especially as your operation expands.

## Drainage

Drainage is key to staving of root drowning, mold, and rot. Make sure when potting that any drainage holes in the base of the pot are sufficiently sized and clear of obstruction. A layer of sand or fine gravel along the bottom of the pot can also encourage drainage.

## Soil

While soil isn't really equipment, it does merit discussion. Not all soil is created equal, and its composition will change over time as your plants deplete its nutrients. Any premixed potting soil is usually fine for growing cannabis and more often than not will not require any changes. Any changes that may be necessary won't be apparent until the plants have been growing for some time.

It is also important to note that although an appropriate pH level is a necessary part of the healthy balance that makes productive plants, fiddling with pH levels through the addition of soil additives, if done incorrectly, can do much more harm than good. The information contained in this segment is important, but exercise caution when adjusting pH levels and keep in mind that adjustments should only be made as needed.

## pH

Remember the pH scale from Earth science class? Neutral is labeled as 7. Solutions with a value less than 7 are acidic, while solutions with values higher than 7 are alkaline or basic. Battery acid has a pH of 1.0, while household bleach has a pH of 12, making it highly alkaline. Soil is either acidic or basic based on its composition, and contrary to what you might think, optimal growing soil is very rarely pH neutral.

The optimal soil pH for growing cannabis is 6.0-6.5, values that are considered slightly acidic. What this means for the plant is that the nutrients it needs are water soluble or can be dissolved and readily absorbed. If the soil pH changes drastically, then your plants may not be able to absorb nutrients; even if they are abundant in the soil, they won't dissolve into water and the plant cannot absorb and utilize them. The pH of soil can be tested either using test strips or with pH meters, both of which should be available in your local home and garden center. To avoid drastic changes to your plants' environment, think preventatively. Test the pH of soil before planting so that young plants aren't killed if adjustments are necessary. Test frequently during the growing process to ensure that the soil pH remains at acceptably acidic levels.

If the soil pH tests too high or too low, then corrective action is necessary. Lime, a calcium based chemical, can be added to the soil to increase the soil's pH—making it more basic—as per the instructions on the package. If the soil requires a lower pH—making it more acidic—then sulfur should be added, again, as per the instructions printed on the package. When handling soil additives known as amendments, exercise caution and follow all safety precautions listed by the manufacturer.

Keep in mind that soil amendments are broken into two categories based on how quickly they change the pH of the soil. Slow release amendments change the soil slowly over time, while the quick release option has a more immediate effect on soil pH. When adjusting pH it is always best to think preventatively. Slower release is almost always the right choice for your plants because it prevents drastic environmental changes that could stress your crop. Test often to stay ahead of depletion, and if any soil amendments are added, test again to ensure effectiveness.

## $CO_2$

All plants use carbon dioxide as part of their respiration; they take in $CO_2$ and expel oxygen. For us mammals this is fantastic; you can't tell me you don't like breathing oxygen. The problem arises when you look at the environment of your grow room or area. If you or I were locked in a sealed room, eventually we would use up all of the oxygen—the air in the room would be replaced with the carbon dioxide we would exhale. The same is true for your cannabis plants, except in the inverse. As your plants 'breathe' in the grow room, they use up all of the $CO_2$ and replace it with oxygen.

Hanging out in your grow room is actually good for you (other than it being a little hot and humid in many cases). The air smells fragrantly of marijuana and each plant is pumping out clean oxygen, which can give you a clear mind. When it comes to your plants, however, the space must be ventilated to change and refresh the air for your plants. For very small operations this isn't too much of a concern. If you open the door to a closet periodically, fresh $CO_2$ comes rushing in and the oxygen will come pouring out.

**For most small operations there will naturally be enough carbon dioxide in the air to sustain your plants.**

For those with bigger operations it will be necessary to install a fan, and the higher the better. Warm air rises, so a ventilation fan that is put toward the ceiling will create circulation throughout the room that draws the stale air up and away from the plants and replaces it with fresh air. Keep concealment and discretion in mind when planning your ventilation. An operation certainly couldn't be called 'stealth' if it is pumping air that is heavy with the scent of fresh marijuana throughout the neighborhood.

In instances in which ventilation already exists to alleviate excess heat, these systems may serve a dual purpose of refreshing the air for your grow room. There are also CO2 enrichment systems available for professional growers that effectively beef up the amount of available carbon dioxide to increase the effectiveness of photosynthesis for your plants and therefore their yield.

## Fertilizer

Fertilizer ensures strong and healthy plants, but too much can do more harm than good. Organic fertilizers are always recommended over chemical fertilizers as the final product will be intended for human consumption. Users agree that organically fertilized buds taste better and create a smoother experience than buds from chemically fertilized cannabis plants. If you do elect to use chemical fertilizers, follow all of the safety precautions provided by the manufacturer.

Different fertilizers will have different specific application, so read the included instructions of use carefully when applying fertilizer. If plants in your crop suddenly start radically changing color and the leaves begin to twist, fold, or curl, these can all be signs of over fertilization. If caught quickly this can be remedied through a process called leaching. To leach your plants thoroughly soak and rinse off all soil to remove the over-fertilized growing medium. Rinse the plants off to prevent runoff from the foliage getting back into the soil. This process should be repeated one to two times a day for three to five days or until symptoms improve. Remember that with the soil cleaned thoroughly, all of the nutrients will be depleted. Quick replenishment methods exist, but for novices it is recommended to gently replace the soil with fresh potting soil.

Whenever you use fertilizer always introduce it slowly and according the manufacturer's recommendations. Check the Appendix at the end of this book for symptoms of common deficiencies.

## Hydroponics

Hydroponics is a subset of the science of hydroculture in which normally land-based plants are cultivated in a nutrient and mineral rich solution instead of soil. Often an inert medium such as gravel is included to anchor the plants in the solution, though solution-only methods exist.

Hydroponic cultivation of plants is by far the most efficient and effective method of growth and can yield plants that would not have been possible using traditional soil cultivation. There is of course a catch: hydroponic systems are complex and require lots of care. As such it is not recommended that new growers attempt to get started with a hydroponic system. While many people professionally grow cannabis using advanced hydroponics, the level of technical involvement is high and requires not only a significant time investment but a monetary one as well. The specifics of cannabis cultivation using a hydroponic system are outside the scope of this book.

# CHAPTER SIX
## STRAIN SELECTION

By now you know that not all types of cannabis are created equal. Different strains have different growing characteristics, different needs, and different qualities for the end user be they medical or recreational. This chapter focuses on the information that new or beginning growers will need to get their first harvests.

If you have the luxury of being able to select a suitable grow room or outdoor grow area from a number of options, then congratulations: you'll have a lot more options when it comes to strain selection. The bulk of growers grow indoors, and as such have additional challenges when growing strains that are bred to be taller. Though these plants can be trained to be a better fit for the grower's space, it is much simpler to start with plants that have been bred to grow shorter and bushier. In many cases these strains have been bred for their size characteristics without sacrificing yield.

## Genetic Engineering

The media-hyped term GMO (genetically modified organism) has lent quite a bit of misunderstanding to the process of genetic modification. While the long-term effects of scientific procedures that tamper with the DNA of various food crops have not been fully studied, this is just one facet of genetic modification.

Humanity has practiced selective breeding for thousands of years; the best example of highly successful and differentiated selective breeding is man's best friend. Think about the variety of domestic dogs that exist. Almost none of those breeds would have naturally come about. Through selective breeding the species was shaped and culled in such a way that designer breeds as well as dogs bred for herding or security are commonplace. The same concept applies to plants. 'Modern grains' differ from the increasingly popular 'ancient grains' because thousands of years of cultivation have resulted in wheat, barley, and rye that are highly distinguishable from the 'natural' make up of their ancestors. Bred for disease resistance, increased yield, lower nutritive needs, and hardiness, these plants never would have naturally come about. Breeding through sexual reproduction combines genetic material from a set of parents from dogs, to cannabis plants, to orange trees. Tampering with the natural process is a form of genetic modification but one that follows the template laid out by nature and produces more or less natural offspring.

Due to the high profit incentive that cannabis offers, scientists, growers, and botanists alike have explored the cannabis genome and have bred a number of strains that are optimized for various characteristics. Today, with legalization on the horizon, that research is expanding and gaining steam to the extent that nearly all seeds purchased are the result of extensive genetic engineering by way of selective breeding.

## Seeds

Now that we have all of the supporting aspects of your growing operation squared away we've come to the good part: the plants. No matter how large or small your operation, you will need seeds. A good source of seeds is a breeder that you know and trust, a resource that can be difficult to come across when just starting out. If personal contacts aren't an option, there are a number of reputable online dealers as well, though it is advised that you thoroughly research the legality of purchasing seeds online, the legality of their transport and storage, and the consequences of violating federal, state, and local laws.

Also, keep in mind that logging onto a marijuana chatroom and posting "Looking 4 weed seeds!!" shouldn't fit into your security profile. Be cautious and casual, and remember that many potential sellers may be equally cautious. It is important to understand: they don't know you and they stand to lose just as much if not more if you turn out to have intentions other than what you state. In the end, your goal should be to establish a rapport with a reliable and trustworthy seller to 'keep you in the seeds' so to speak. Any successful business needs a stable supply chain. It is advisable, however, to use online forums and cannabis chatrooms where amateur and professional growers alike can swap stories, tips, and tricks. In a tightly knit community such as cannabis growers, a bad breeder's reputation will precede him. The same goes for a successful and trustworthy breeder; people will be willing to review and provide references for a breeder and seller with a proven track record of integrity. It is common knowledge, as with many industries with varying legality, that the Internet is rife with scams advertising discount seeds.

Don't abandon online communities once you have found your source of seeds—forums, message boards, and chatrooms can all be valuable resources as your operation grows.

In some states, the sale of cannabis seeds is legal and can be done over the counter at a physical shop. In that instance, don't worry about the process. Starting your marijuana crop will be as easy as a trip to the hardware store. Staff at such stores are invariably knowledgeable and helpful and can help you tailor your strain to your situation.

## Understanding Different Strains

The wonderful world of growing cannabis is packed to the gills with different kinds of marijuana. There are literally hundreds of different strains, but to narrow our focus a little bit, we're going to look at which strains are best for beginners.

Some online retailers will break their offerings into different categories to simplify the selection process. This can include categorizing strains by climate zone[2], height, flowering time, yield, and sativa or indica. Remember in the very beginning when we discussed the different types of cannabis, C. sativa and C. indica? That's going to come back into play here. Understanding if your strain is sativa or indica-based is important, as this will tell you some fundamental characteristics to expect from the growing cycle and the final product.

---

[2] Climate type and growing zone aren't important when growing indoors, but outdoor growers should look up their growing zones here at http://planthardiness.ars.usda.gov/PHZMWeb/. This USDA-provided data can help you select a strain that you intend to grow outdoors.

| DIFFERENCES BETWEEN SATIVA & INDICA ||
| Sativa | Indica |
| --- | --- |
| Visually distinguished by thinner and more slender leaves. Sativa strains grow taller with a more loose branching pattern | Visually distinguished by broader leaves. Indica strains grow shorter and with a much denser branch pattern |
| From more equatorial regions. Taller plants with a longer growth time before flowering | From mountainous areas with harsher climates. This means a much higher resin production and a shorter flowering time |
| Best for outdoor growing | Best for indoor growing |
| Approximate reported effect: appropriate for daily use, energizing and uplifting, spacey and hallucinogenic | Approximate reported effect: appropriate for use at night, calming and relaxing, body high or buzz |
| Commonly prescribed to treat: depression, ADD, fatigue, and mood disorders | Commonly prescribed to treat: anxiety, insomnia, pain, and muscle spasms |

*Table 8*

There is a third classification of hybrids, or plants that have been bred to be a combination of both sativa and indica plants. Often these strains are bred with an intended effect in mind or a particular growing characteristic. Your seed dealer can tell you more about the particular characteristics of different hybrid strains.

## Auto-flowering Seeds

When shopping for seeds, in addition to the standard lots of seeds you may see two additional categories: auto-flowering and feminized seed lots. Auto-flowering seeds are excellent for novice growers as they are easiest to care for. When growing auto-flowering seeds there is no need to remove the male plants or any need to change light cycles.

Auto-flowering cannabis plants are hybrids that incorporate the uncommon and little known cannabis ruderalis, a cannabis variant that grows wildly in the Caucasus Mountains, Mongolia, and China. C. ruderalis flowers automatically, no matter the change in light, and spring up very quickly. It is also a very short plant that is a dwarf next to even C. indica. The downside to this small cannabis variant is that in its natural form it produces low amounts of THC giving it poor utility. When properly crossed with sativa or indica cannabis plants, however, the benefits of both types can be reaped and the result is an easy-to-care-for, quick growing, and frequently yielding cannabis plant.

Many growers swear by auto-flowering strains, and they often present the best option for new growers. Some growers who started with auto-flowering strains are still growing them today, never switching to pure indica or sativa strains. Growers agree that the quality of buds from auto-flowering strains can be optimized to the point that many users can't tell the difference. Understandably this type of strain is quickly gaining both popularity and scrutiny from the growing community.

## Feminized Seeds

Feminized seeds have been bred to contain no male chromosomes, therefore ensuring that every crop will be only composed of female plants. This option is also a good choice for new growers, but keep in mind that these plants should not be used for breeding or breeding projects. The process that is used to 'feminize' the plants comes from a number of self-pollinations and various other methods that destabilize the plant's DNA. Growers often report seeing a higher number of buds that display both male and female traits—hermaphroditic buds—with the offspring of feminized plants.

## Name That Strain

Many seed dealers offer strains that they recommend for new growers with a variety of names. Once you have found a reputable seed dealer, work with her to find a strain that is a good fit for you, and use the information in this chapter to help make your determination. Alternatively, if you know the name of a strain that you like, look up its characteristics and determine if it is a good fit for your operation and your needs.

A variety of websites and other outlets have extensive amounts of information regarding the individual effects that various strains have on the end user as well as optimal growing conditions and other tips and tricks. The massive number of strains that exist, and the new ones that are being added all of the time, means that we can't include a comprehensive list here. Work with your breeder or seed dealer armed with the knowledge this book has given you to make the best kind of decision: an informed one.

# CHAPTER SEVEN
## GROWING & CARING FOR YOUR CANNABIS

This is it, the step-by-step guide to harvesting your first crop. If this doesn't turn out the way you were expecting, be patient and try again. Learning is a process, and becoming an experienced grower means that you need time to gain that experience. Remember: record everything! Keeping a journal or a grow log will ensure that you don't make the same mistakes twice and that the right decisions are duplicated next time. Having a recipe for success will help others help you as well. If you seek help from more experienced growers, whether in person or through an online forum, keep your log handy to help them understand what you have already done and possibly where you may have gone right or done wrong.

## Growing Plan

The following schedule and growing plan covers the whole process from seeds to mature, flowering plants.

### Germination

The first stage of a plant's life is germination. For cannabis plants germination can take up to three weeks, and the seeds will require daily attention. The process of germination is the emergence of a root from the shell of the seed. In nature the seeds drop and germinate on their own in exposed soil, but for a grower's purposes it makes more sense to stimulate germination indoors. Scatter the seeds on a wet piece of paper towel and place them on a plate. Cover the seeds with another wet paper towel, then place another plate on top of the first. The paper towels should remain damp; check twice a day to ensure they are suitably moist.

Once the shell is cracked and the new root is approximately an eighth of an inch long, your seeds can be transferred to a growing medium. Do not plant the young seeds any deeper than an eighth of an inch. Exercise extreme caution when transferring the germinating seeds, it is best to gently dump the entire damp paper towel into the growing medium and not to touch or otherwise directly disturb the sprouting seeds.

A quick note regarding your growing medium: there are a variety of commercially available growing mediums ranging from standard potting soil to coco fiber. Some growers use small trays of temporary pots to start the germinating seeds then later transfer them to larger, more permanent, homes. Others simply start germinating the seeds in their intended homes, whether in pots, trays, or outdoors. Experiment with which works best for your situation and operation.

Once your seedlings have grown several leaves, it will be safe to transplant them into a rich potting soil mix. It is also at this stage that they will be hardy enough to take in fertilizer. It may be enticing to grow the largest plants you or anyone has ever seen, but before you start making calls to brag, keep concealment and security in mind. Many smaller strains have been optimized to produce a comparable yield, so there's no trade off to growing smaller plants. Smaller plants will be easier to conceal, easier to transport, and easier to care for.

Always use extreme caution when fertilizing your plants, especially when they are very young. Too much fertilizer can produce an effect opposite to your intent: too much can kill young plants. When using chemical fertilizers, take care to read all of the directions for safe and proper use. Fertilize with care and remember that under-fertilization is better than over-fertilization.

If your operation is small and you intend to remain more or less small, keep in mind that plants, like goldfish kept in small bowls, will only grow as large as their pots allow them to. Keeping your cannabis plants in smaller pots means that the plants will stay small, concealable, and portable. Perhaps you have heard of small recreational operations whose entire crops are grown and harvested from red Solo cups; it's more than an anecdote—it's very possible, just don't expect record breaking yields.

## Week 1: Vegetative

Your established seedlings are in the soil and ready to start growing. Remember not to shock them (young or otherwise) with sudden changes. This first week will be acclimating your plants to the intensity of the growing lights. Run them at 50%, either turn them to half strength or only use half as many lights. This is part of the vegetative growth stage, so remember blue spectrum-rich light will give you the best results. For this example assume that 600 watt HID lights are being used.

When watering for the first time, ensure that they are off to a good start by watering until excess water flows from the bottom of each pot. This will be more than enough water for a while, so wait to water them again. It is recommended to spray them daily to keep the humidity up; this will help reduce moisture evaporation from the soil as well.

If you are using higher-intensity lamps, make sure that they are far away from the young plants. About 3 feet is a good distance for plants this young. The light duration for this period is 18 hours on, 6 hours off.

The planting density here is one plant per square foot.

## Week 2: Vegetative

During week two turn on (or up) all of the lights and reduce their distance from the plants by half. This should bring them to 1.5 feet from the young cannabis plants. If there is any concern about excessive heat or the plants becoming too hot, then back the lights off a bit. If possible, water your plants from below by dipping them in water to train the roots to seek the bottom of the pot faster. Do not leave your pots in standing water. Water normally as needed. This will expand the root system quickly and ensure that your plants are getting the most out of their watering regimen.

To train your plants to grow firm stems, set a fan to the lowest setting and point it at your crop. The plants should move only slightly under the breeze of the fan. As the stalks bend and move slightly, the plant will divert resources to strengthen its core. Thick, strong stems mean that your plants will be hardier. Hardy plants better survive transport, more effectively transport nutrients throughout their own systems, and have a higher tolerance to stressors like under-watering. Most importantly, thick stems prepare the plant for the weight of thick and heavy buds.

During the second week of vegetative growth continue the 18 hours on, 6 hours off light cycle. In many instances the profit-motivated grower can switch to the 12 hours on, 12 hours off cycle to stimulate flowering after only two weeks. If you continue another week of vegetative growth, it won't hurt your plants, but in most cases growers often make the switch after two weeks if their plants are strong and hardy enough. When starting out it may be wise to go at least one more week before changing to the flowering light cycle, but that is up to you. Remember to record everything when determining what works best for you.

## Week 1: Flowering

Here your lights should still remain at 100% capacity, and you should make the shift to your red spectrum. Here is the place for your HPS lights if you have taken the plunge and purchased those. Your lights should remain about a foot and half away from the plants if they are high-intensity.

Now you make the change from 18 hours on, 6 hours off to 12 hours on, 12 hours off. This effectively tricks the plants into responding to what would normally be autumn light and day length. Because red spectrum light is effectively hotter, it is a good idea to aim your fan between the plants and the lights to diffuse the heat throughout the entire room and combat hot spots that could fry individual plants. Keep strong breezes off the plants.

Keep in mind that although this begins the flowering process, you will not see signs of flowering for a few weeks yet.

## Week 2: Flowering

Using the same lighting setup and light cycle, keep an eye on the tops of the plants and the edges of the leaves. The plants should be growing noticeably faster now, so monitor their growth to ensure that they are kept a safe distance (about a foot and half) from HID lights. If the edges of the leaves begin to curl or brown, the lights need to be farther away or your fan needs to be turned up and repositioned.

It is also a good idea to watch the base and stem to 'pinch out' side shoots or other small plants that may be developing on your main plant. Nonproductive parts of the plant drain resources and don't contribute to bud production.

## Week 3: Flowering

Again, your lighting shouldn't have changed except for any adjustments to stop the plants from growing into them or too close, and the light cycle should still be 12 hours on, 12 hours off. The root system should protrude from the bottom of the pot, and the signs of the first flowers should be visible. This is the last possible time when males and females can be commingled. If there are any males that remain with females in your crop, you risk your entire harvest.

During this period the plants consume the highest amount of water.

Check moisture levels regularly and water as needed. If you are using a fertilizer, skip this week and rinse the soil.

### Week 4: Flowering

At this point small, developing buds should be present all over the plants, and the plants should start to become fragrant. For those growers using fertilizer, switch to a bloom fertilizer during this week. Remember to make sure that as the plants grow the lights are kept at a safe distance from the plants as they get taller. Throughout this week the lighting cycle does not change.

### Week 5: Flowering

At this point your crop should slow its upward growth and the buds should develop more readily. Here light is key, but remember that the buds are even more sensitive than the leaves, so make sure that nothing gets cooked. If you have concerns about heat, a simple thermometer stuck in one of the pots can tell you specifically how much heat the plants are experiencing. Keep an eye out for signs of distress, disease, or insects. If any of your stems are red, this can also be an indicator of stress.

### Week 6: Flowering

At this point the plants require the maximum amount of water and $CO_2$ so check the moisture frequently and make sure that your plants have fresh air. This means keeping a ventilation fan running. Right now your job as a grower is to make sure that your plants can do their thing in a stable and optimal environment. If you're using fertilizer, rinse at the end of this week.

### Week 7: Flowering

The buds that are present are gaining volume and there is a white material deposited on the small leaves near the buds. This is an indicator of high THC levels, so keep up the good work.

### Week 8: Flowering

This is the time when the buds will become denser. If the bottom leaves of your plants are becoming discolored and dying, it's not a nutrient deficiency: it's a natural part of the flowering process, so don't over-fertilize thinking the plant is dying. Keep an eye out for mold or upper leaves that are yellowing or brown. If anything appears to be moldy or otherwise damaged, clip it immediately to prevent it from spreading.

## Week 9: Flowering

Continue to keep an eye out for rot or insects. During this phase it is normal for many of the leaves to become discolored, and the hairs on the buds should be turning brown. Once at least 80% of the hairs have turned brown, they are ready to harvest. If they are still using high amounts of water, then they can be left a while longer before harvesting.

Please note that it is important not to fertilize your plants during the final stages of flowering. Additional fertilizer at this stage can reduce potency and can add unwanted bitter flavors to the final product.

And there you have it! Remember to record everything about the process so that it can be reliably replicated and so that you have material to bring to others if questions arise, and look back through this book frequently to keep all of the information in the forefront of your mind when growing.

# **CONCLUSION**

While growing cannabis today may be a very different process than it was 5000 years ago, the end result is the same. When you undertake the cultivation of cannabis, you are embarking on not only a profitable business venture but a time honored tradition that has been with humanity for generations upon generations. Whether you are growing an empire of green or a closet operation that keeps you supplied, the cultivation and enjoyment of cannabis (or the money it brings) can be a very rewarding experience.

Never before has the time been better. With the United States on the brink of enabling legislation that will make those who are poised to leverage the expansive and untapped market rich, the incentive for individual farmers, entrepreneurs, and aficionados to learn more about the wonderful world of growing cannabis is increasingly real.

Don't hesitate, better buds are within your reach!

# APPENDIX
## SYMPTOMS OF COMMON NUTRIENT DEFICIENCIES

### Calcium
A calcium deficiency is evidenced by small brown spots or dead areas in the leaves. The entire leaf structure also becomes crinkly and brittle to the touch. The easiest and earliest indicator of a calcium deficiency is abnormal smallness in new growth. New growth may also exhibit shape distortion and curled tips. Calcium deficiency is easily corrected with lime, or a store-bought calcium supplement, most often accompanied with iron and magnesium. As always, be sure that the soil PH is adequate, otherwise no matter the corrective measure you implement the roots will not be able to absorb the soil supplement.

### Boron
Boron deficiencies can be identified by yellow-colored new growth, abnormally heavy or thick growth tips, and rough or hollow stems. New growth may look scorched or burnt, and this can be distinguished from heat or light damage by comparing the plant in question to those around it with similar lighting conditions. In order to correct a deficiency of boron, ensure that your plants are adequately watered, that they have adequate nitrogen and potassium, and that the PH level of the soil is correct.

### Copper
A copper deficiency is demonstrated by unusual coloring and curling back of a plant's leaves. Newer leaves may appear to be twisted and dark colored while older leaves appear lighter or yellowish. In the case of a copper deficiency, it is best to flush your soil with pH 7 water mixed with a half-strength dose of organic fertilizer. Leaves that become yellow in between the veins of the leaves with brown areas may indicate another nutrient deficiency. These other symptoms may be accompanied by stunted overall growth of the plant. The solution to this problem is essentially the same as that for a copper deficiency: flush the soil with PH 7 water that has been fortified with a half-strength measure of organic fertilizer. The number one preventative measure for

mineral or nutrient deficiencies is to make certain that you test your soil PH daily and keep it within the acceptable range for cannabis. Many nutrient deficiencies can be avoided with a proper soil pH that allows your plants to correctly absorb water soluble nutrients.

## Over-Fertilization

'Nutrient burn' or nitrogen toxicity is a common symptom of over-fertilization. This is evident by browning or yellowing tips of foliage that spreads to the entire leaf. In the case of nitrogen toxicity, leaves may show signs of clawing—a condition that causes leaves to fold upward. Some foliage may be a darker green while some may turn yellow or brown on the tips. Nitrogen toxicity is often mistaken for over-watering, but diligent monitoring of soil moisture levels should rule that out as a cause. Any over-fertilization issues can be quickly resolved by leaching the soil with a half-strength solution of organic fertilizer 1 to 2 times daily for 3 to 5 days. Pay very close attention to your crop as harvest time approaches. The foliage naturally turns yellow and shows signs of a nitrogen deficiency as buds begin to grow and form. It is best not to fertilize during this final period.

## Light & Heat Damage

Additional common problems with your crop may be the result of light or heat burning. Light or heat burning is most evident in the leaves that are most directly exposed to the light or heat source, so pay attention to the pattern of damage or damage symptoms. Leaves that are closest to the light or heat source may show signs of stress such as yellowing or browning and may actually appear burnt altogether. Excess heat causes the edges of leaves to turn up or curl and may cause the entire plant to wilt. These problems are easily corrected by immediately lowering the temperature of the growing environment, ensuring proper moisture content in the soil, increasing air circulation and moving light and heat sources away from the affected plants.

If this happens to your growing operation, re-evaluate your heat sources and light sources and make permanent corrections as needed. It is also important to regularly check the distance that your lights are from the tops of your plants. That distance changes frequently as your plants grow closer to the light, so a lapse in vigilance can result in some cooked cannabis.

# ABOUT
## CLYDEBANK ALTERNATIVE

ClydeBank Alternative is a division of the multimedia publishing firm ClydeBank Media LLC. ClydeBank Media's goal is to provide affordable, accessible information to a global market through different forms of media such as eBooks, paperback books and audio books. Company divisions are based on subject matter, each consisting of a dedicated team of researchers, writers, editors and designers.

The Alternative division of ClydeBank Media is composed of contributors who are experts in their given disciplines. Contributors originate from diverse areas of the world to guarantee the presented information fosters a global perspective. Contributors have multiple years of experience in horticulture, alternative medicine & healing.

For more information, please contact :
info@clydebankmedia.com

# MORE BY
## CLYDEBANK ALTERNATIVE

**Cannabis Oil Essentials:**
A Comprehensive & Enhanced Guide
To Cannabis Oil Benefits
Visit : http://bit.ly/cannabisessential

**Companion Planting For Beginners:**
Simple Ways To Dramatically Increase
Crop Productivity With Companion Planting
Visit : http://bit.ly/comp_planting

**Growing Organic Berries:**
Exactly How To Grow, Maintain & Preserve
Every Type Of Berry To Support A Healthy Lifestyle
Visit : http://bit.ly/growing_berries

# GET A FREE CLYDEBANK MEDIA AUDIOBOOK
# + 30 DAY FREE TRIAL TO AUDIBLE.COM

## GET TITLES LIKE THIS ABSOLUTELY FREE:

- Business Plan Writing Guide
- ITIL for Beginners
- Stock Options for Beginners
- Scrum Quickstart Guide
- Project Management for Beginners
- 3D Printing Business

- LLC Quickstart Guide
- Lean Six Sigma Quickstart Guide
- Growing Marijuana for Beginners
- Social Security Simplified
- Medicare Simplified
- and more!

## TO SIGN UP & GET YOUR FREE AUDIOBOOK, VISIT:
www.clydebankmedia.com/audible-trial

Made in the USA
San Bernardino, CA
23 May 2017